Amazing Magnets!

by Caroline Hutchinson

We see magnets in many shapes and sizes. We use magnets in many ways.

How do magnets work?

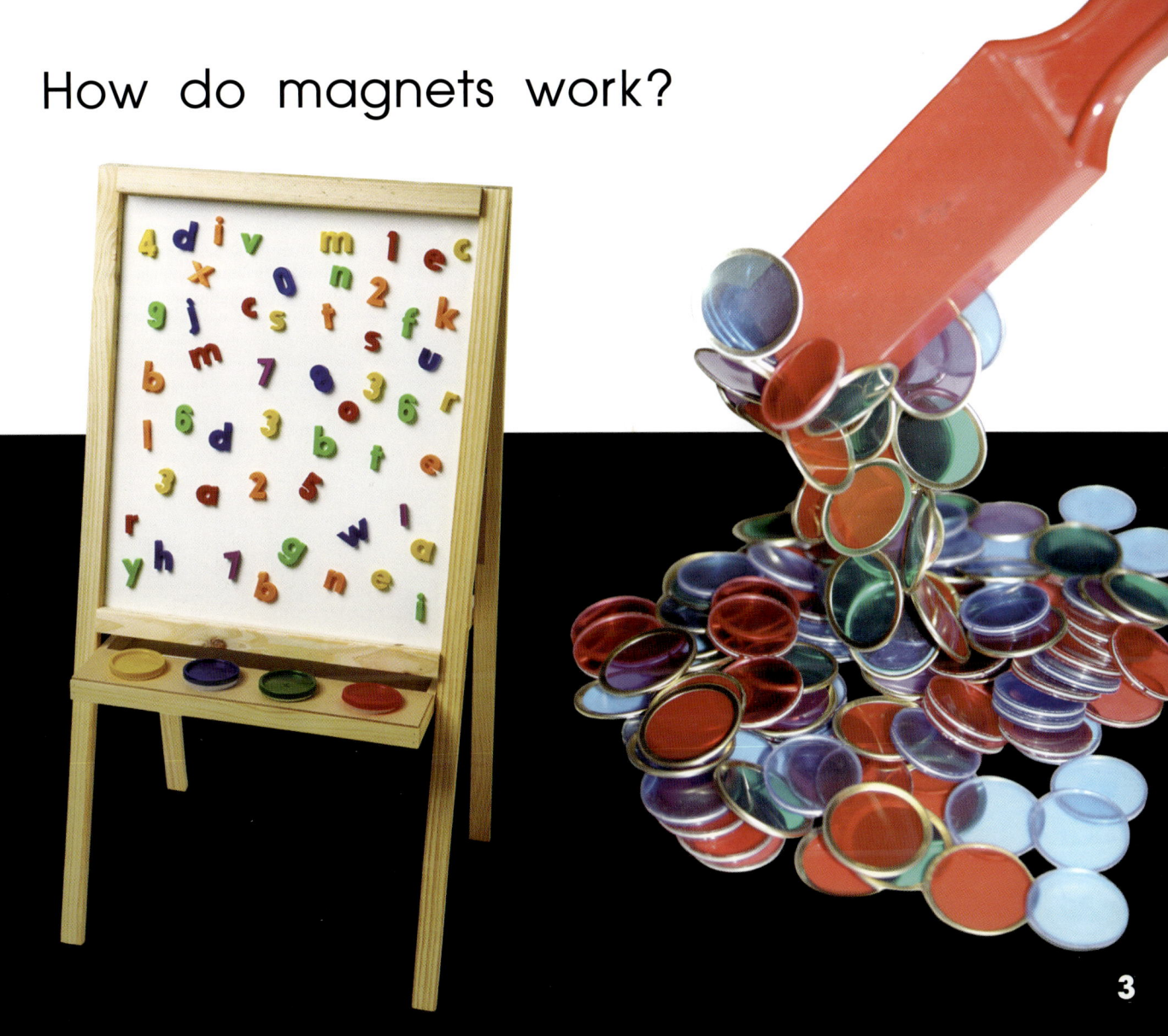

A magnet pulls some objects.
First, find some objects that
a magnet can pull.

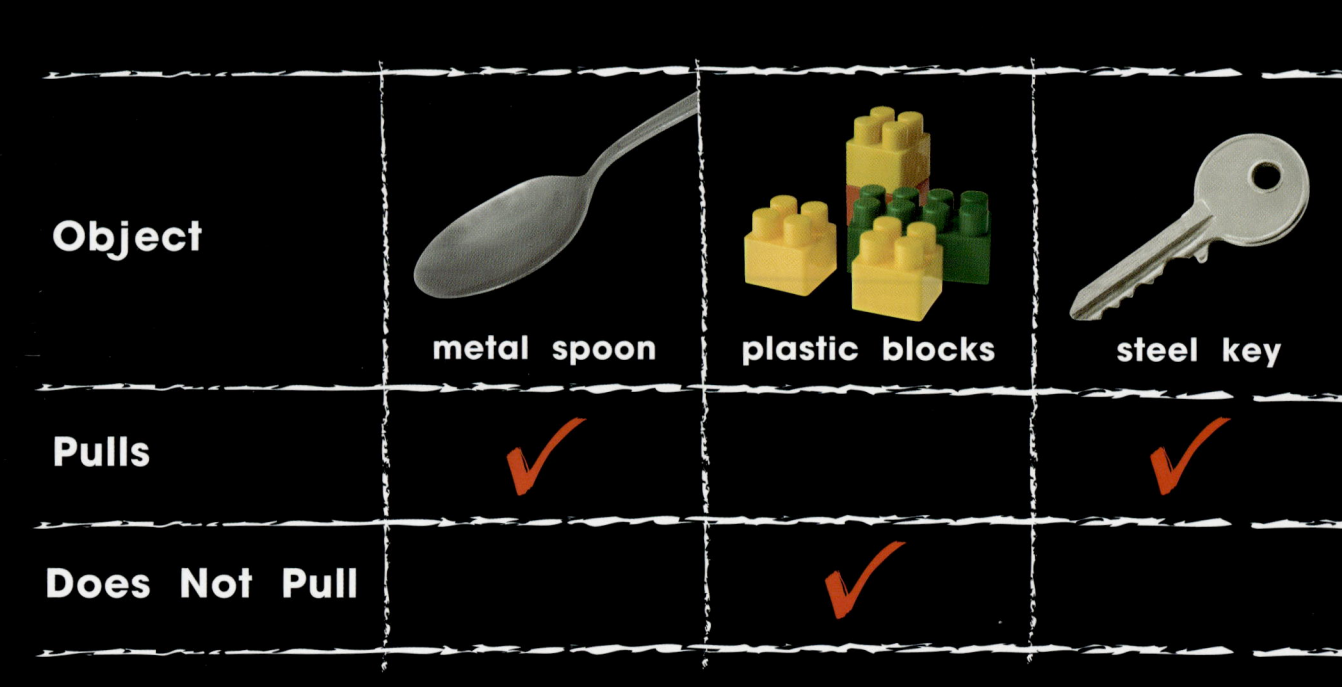

Object	metal spoon	plastic blocks	steel key
Pulls	✔		✔
Does Not Pull		✔	

▲ The chart shows objects a magnet can pull.

Then find some objects that
a magnet does not pull.

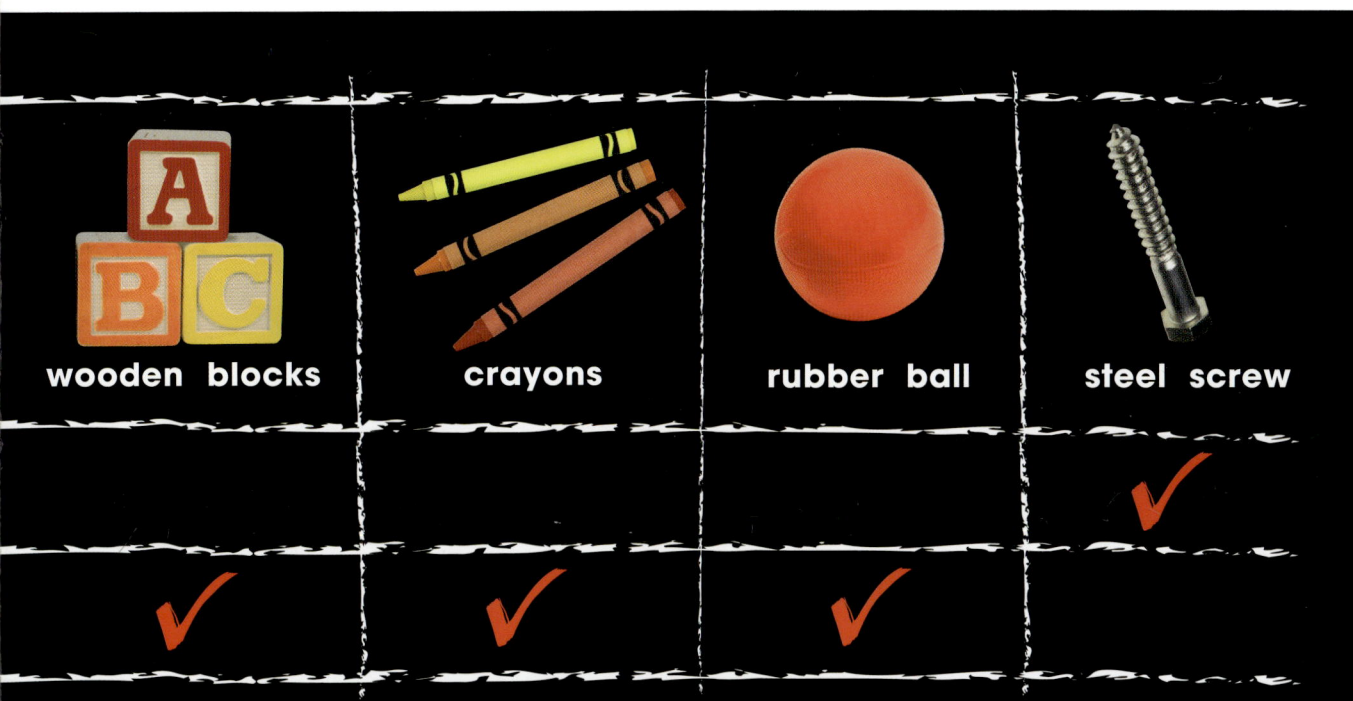

wooden blocks	crayons	rubber ball	steel screw
			✓
✓	✓	✓	

▲ The chart also shows objects a magnet cannot pull.

Some magnets pull
other magnets.

▲ These magnets attract.

Some magnets push
other magnets.

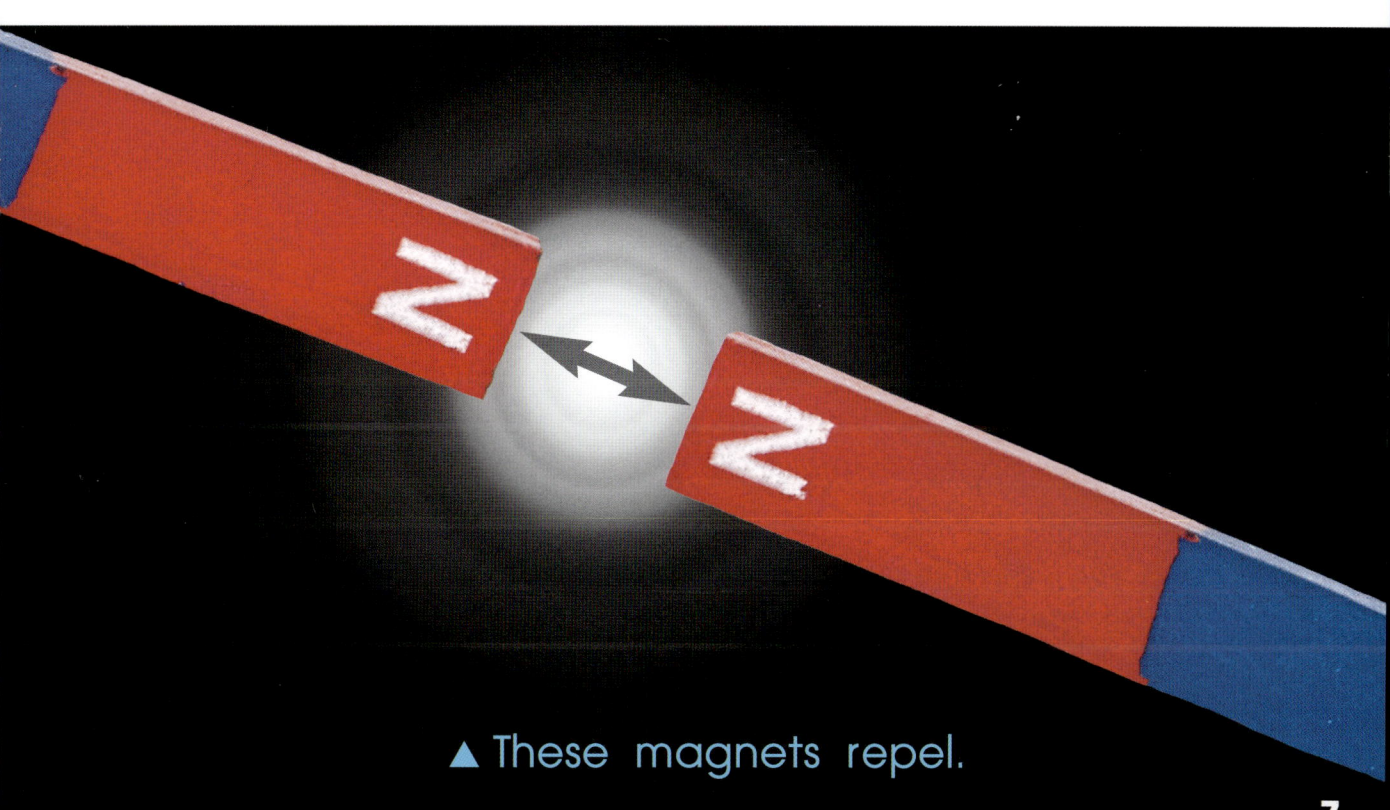

▲ These magnets repel.

Some trains use magnets.

This train has magnets that keep the train on its tracks.

▼ These magnets are very strong.

We can use a magnet to make a paper clip dance!

Gather these materials. ▼

First, clean the magnet
and the paper clip.

▲ Use a paper towel.

Next, cut the string. Tie an end of the string to the paper clip.

Then tape the other end
of the string to your desk.

◀ Tie the knot tightly.

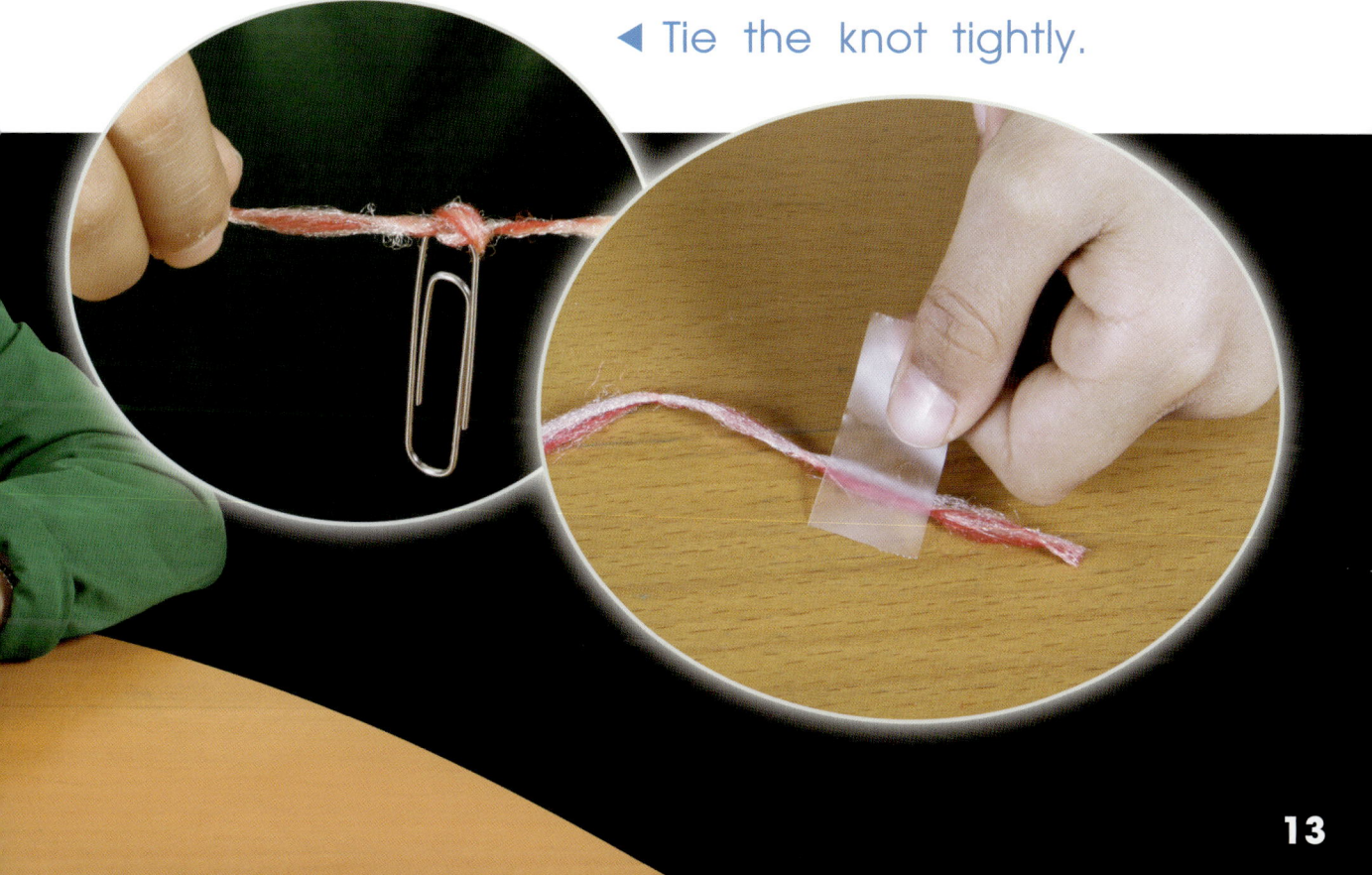

Now, bring the magnet close to the paper clip. Use the magnet to pull the paper clip up.

▼ Slowly lift your hand.

Then move the magnet back and forth. Watch the paper clip dance!

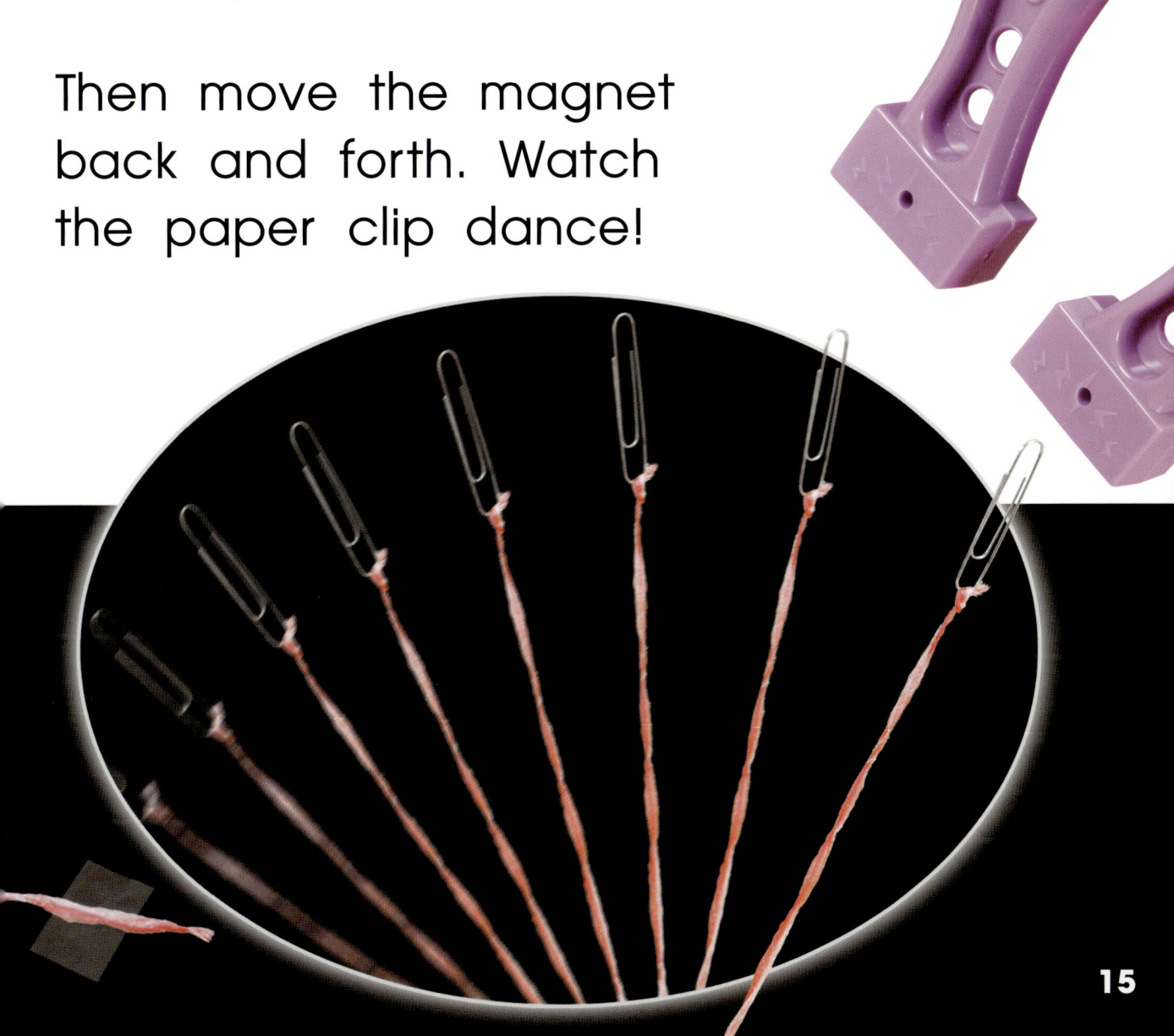

We see many types of magnets.

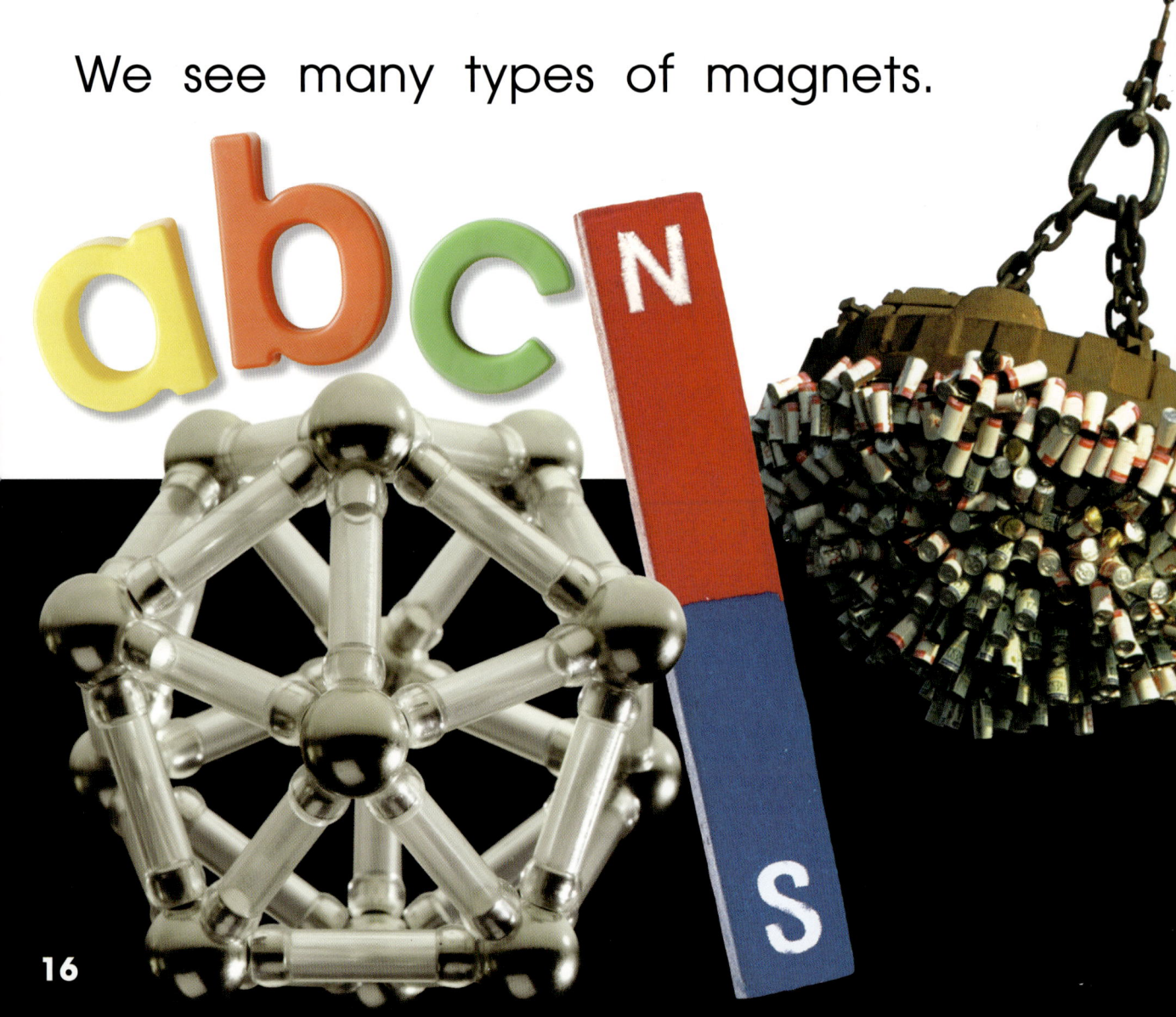